Rebel Science

Rebel Science

An eye-popping glimpse of the universe - and a few things much closer to home

Nat Krinsen

Be inspired by the scientific adventure in an hour

Rebel Science
Copyright © 2012 by Nat Krinsen

Published by Fovea Publications

fovea@live.co.uk

"Minds are like parachutes. They only function when they are open."

Sir James Dewar, Scientist (1877-1925)

Men love to wonder, and that is the seed of science.

Ralph Waldo Emerson

'The universe is not only queerer than we suppose; it is queerer than we can suppose.'

J. B. S. Haldane, Biologist

Fed Up?

If I got your favorite chocolate bar and *made* you eat it continually, bar after bar, not allowing you to stop, in the end force-feeding you, you would soon come to hate it (and me!). That's what school did with feeding you science. You've had it stuffed down your throat.

Once, when you were tiny, you used to ask questions, driving your parents mad with a continual "why?" When you were young everything made you ask "why?" Everything interested you. Any new thing was a fresh puzzle that demanded yet another "why?" Can you just about remember those distant days when the world around you was full of fascination and raised endless questions? Do you remember the thrill of seeing your first dinosaur picture, or of a space ship, or some giant squid of the deep? It stirred something deep within you. You wanted to know more. Everything you saw was part of an undiscovered world full of adventure – and endless mystery. You longed to know more – and yet more.

And then they started teaching you science at school, ramming facts down your throat . . . and the lights of your mind started going out one by one. You stopped asking questions; there was too much dull "stuff" to understand and remember. And then came the final killjoy, some utterly pointless homework was demanded from you! You stopped asking questions; you stopped thinking. Your mind began to shrivel, to die.

Let me tell you a secret, a rather horrid secret. Science in school is often taught upside down, so it is no wonder it does not make exciting sense.

It is given to you the wrong way round. What you start with are building blocks at school. You are given little bricks, one at a time, so that one day, in the distant future, if you live long enough and don't die of boredom, the whole thing might just look like something intelligible and interesting to you. The trouble is bricks are boring.

When you used to play with *Lego* it was not the bricks that interested you, it was the big picture – the castle or palace you wanted to build, the adventure you wanted to play out in your mind. Now imagine being *made* to stare at *Lego* bricks, to draw them, to measure them, to weigh them. Imagine one week drawing red ones and another week white ones, and so on, and then looking at different shapes of brick. After this you would be told to practice fixing two bricks together and then pulling them apart. You would lose the will to live.

That is not the way a child gets excited to play with *Lego*. Now imagine something totally different. Imagine when you were at the age when you were starting to play with *Lego* and being taken to *Lego-World* – and there you saw a jaw-dropping massive *Lego* spaceship, or dinosaur, or palace, or something like that. When you got home you would become obsessed; you would want more and more bricks. Your imagination had been set on fire. You had been inspired. You would now soon learn all about the individual bricks.

School often teaches science like staring at *Lego* bricks, little fact after little fact. But what we are going to do now is turn this on its head. We are going to be rebels. We are going to look up and see the mind-bending big picture. We are going to let our imaginations catch fire. It's time to rise up and revolt.

Wake Up!

I hope I have caught you in time. I hope some of the lights of your mind are still there. I hope there is still a glimmer left.

What I want to do here is to stop your mind going to sleep. I want to inject something into your brain that will wake you up, blow your mind and change you – forever. I want to make you dizzy with such mind boggling, weird, bizarre, strange (but true) ideas that you won't ever be the same again.

But before we start I want you to go into the bathroom, fill the sink with ice cold water and then plunge your head under it for a few seconds. And then come back.

Don't turn the page until you've done this . . .

No, don't cheat!

Go back and do it.

It's important.

Vital!

What you now feel physically is what I want to make you feel mentally.

Cold . . .

Fresh . . .

Awake . . .

Startled . . .

Invigorated . . .

Ready for something totally new . . .

And perhaps more than a bit dizzy . . .

No, much more than a *bit* dizzy . . .

What you are about to read will make you feel like your brain is melting . . .

Like your mind is reeling, exploding, and turning inside out . . .

Don't say I didn't warn you . . .

Addiction . . . this way!

But there is more.

I also want to warn you.

Seriously.

You will not only wake up – something else will happen to you.

You will be injecting something into your mind that could be dangerous.

You will see everything differently.

Everything.

You will become an addict . . .

Addicted . . .

Hooked . . .

You will start to crave . . .

To thirst . . .

And you will never . . .Never . . . Be satisfied.

Hooked

After reading this very short book everything you see, hear, feel, taste and smell will become something new; they will become clues to the greatest perplexing puzzle of the universe – the search to understand it, to make some kind of sense of it, to come to grips with it. Your life will become a never ending adventure. You will always want to know, to know more and more. Always.

You are about to enter a great detective mystery, and with clues here and clues there you may begin to piece the bits and pieces together and perhaps even be the first person to make sense of it all. No one has yet.

Most questions are still unanswered. Most things are shrouded in mystery. But as we continue, you will become addicted to grasp more – and I warn you, the more you know the more you will crave to know. It is a spiral of addiction. If you don't want this, stop reading now, close the book, walk away – and everything will stay as it is.

But if you keep on reading this then I am now going to inspire a thirst in you that will never go away, a hunger that will never leave you, a craving that will never be satisfied.

I will make you addicted to the desire to know and understand, to peer into dark mysteries and begin to make sense of them.

If you have decided to continue with me, let us begin.

Let us start our journey . . .

Poking Mother Nature

So what exactly is science? Science is a way of poking the universe to make her give up her secrets.

For thousands of years people like you and me never knew the secret method of how to find out how the cosmos works, and so we couldn't grow enough food, we couldn't heal diseases, we couldn't fly. We simply didn't have the key to unlock the hidden treasure chest where the secrets of knowledge are found. We were ignorant and powerless.

And so we made up stories, lots of myths – and we believed in magic and spells and things like that. But of course these didn't work and so we starved, we died of diseases – and we never flew.

We wanted to understand Nature and we wanted to control Nature – that was our dream, and that was why we believed in magic. But then, over the past few hundred years, we ever so gradually started to realize that there was a key to open the treasure chest of knowledge. We began to realize there was a way of poking Mother Nature so that she revealed her secrets to us – and we could, gradually, bit by bit, begin to control her. We now call that key, that way of poking Nature, the Scientific Method.

Science unlocks the secrets of the universe; it reveals the mysteries of life.

It gives understanding; it gives power.

Revolution

Basically, what this means is that we actually came to realize that knowledge is out there in the physical world. And rather than just guessing about it, making up stories about it and thinking things inside our heads about it, we could actually do something to uncover that knowledge.

We realized that if we carefully push and pull things, then by vigilantly noticing *what* happens we can begin to understand *why* it happens. And then, once we have made our guess at this we can test the idea somewhere else, so that we gradually piece the bits of the jigsaw of knowledge together. These activities we call experiments.

Over the years we have got better and better at doing this. We realized that we must only change one thing at a time and keep everything else the same. Our poking became more and more precise; our key became a better and better fit. Nature's treasure chest of secrets was opening.

Sometimes all this poking and prodding was very cruel. Mother Nature needed to be tortured to give up some of her secrets. Early on, those who wanted to understand the body would often dissect an animal while it was alive (without anesthetic!) so that they could see various organs like the heart actually work. At other times human bodies were stolen from graveyards to extract information from them by cutting them open and trying to work out how they worked. There were even grave robbers who made a living by stealing fresh corpses the night after the funeral and then selling them to scientists.

Many of the adventures into uncovering Nature's hidden information were held up because of a lack of equipment like telescopes and microscopes. But gradually, like a snowball rolling down hill, the explorers got more and more information from the world around us.

And so the revolution happened. We unearthed undiscovered laws and hidden principles – and worked out how to use them. We learnt to make carriages that move by themselves, planes that fly in the air, drugs that heal diseases, ships that could explore the bottom of the ocean, ways of talking to people on the other side of the world. We even went to the moon – and we have sent robot spaceships far, far beyond that. Perhaps most exciting, we have learnt how to actually look back in time and see events with our own eyes that happened thousands and thousands of years ago.

Remember the old fairy tales and myths with flying carpets, and wizards that could make explosions and magic potions to heal the sick. Our science is far better than that old magic and myth. To people only a few generations ago our science would be magic! And our "magic" is deeper, darker, brighter and more powerful than they could ever possibly imagine: we now know how to destroy our whole planet. And we now know how to save it too.

More than all this, our getting and using such knowledge are happening faster and faster. The snowball is accelerating. At the moment our total scientific knowledge is doubling every few years – and even more incredibly, the rate at which this is happening is doubling within a decade.

When we look back we see a revolution – and as we look forward . . . well, it is beyond anyone's guess what we will uncover and what we will be able to do. The sky is nowhere near the limit. Nowhere near!

This is utterly incredible.

But it raises a very personal question for you.

That's right, a question for you.

Do you want to be part of this revolution?

Or, are you happy to be left behind?

If you want to join in, read on . . .

What we are going do now is to set out on an adventure.

We are going to enflame our imagination.

We are going to fling open the windows of our minds.

We are going to be part of a revolution that will change the world.

Let us begin . . .

What's the time?

About a hundred years ago a massive shock happened in the scientific world – and we have never recovered from it.

Someone called Albert Einstein did a few rather complex calculations and discovered something that no one has really been able to get their heads around ever since.

What he thought he discovered was that time is not what you think it is. His calculations suggested that time does not run at a fixed rate. Time can slow down and time can speed up.

He calculated that the faster you travel the more time will slow down for you compared to a person who is not travelling so fast.

More than that, he worked out that time also slows down as gravity increases. That is, time runs slower near, say the sun, with its gravity strongly pulling you inwards, rather than far out in space where its gravitational attraction is much, much weaker.

Could that really be true?

Could Nature really be that strange?

Could time really s p e e d up and slow down?

Could the universe be *that* weird?

Surely he got his sums wrong!

He didn't!

Years later, technology had advanced enough to test these strange ideas.

By putting super-accurate atomic clocks in supersonic jets and on orbiting satellites, scientists were stunned to find that Einstein was absolutely right!

In 2010 this amazingly was confirmed using special ultra-accurate optical atomic clocks!

Time really slows down the faster you go. This means that if you went up in a space rocket and travelled much faster than we can now achieve, when you came back to earth everyone would have aged faster than you had. Theoretically, if you travelled fast enough you could come back and find everyone you knew had died and it was, say a hundred years later!

What this means is that, in theory at least, you can travel forward in time – but it is a one-way ticket. You could never come back – not by that method, anyway.

The scientists then tested the effect of gravity on time by using a very accurate method to measure the speed of time passing at the top of a mountain compared with the bottom. This experiment is based on the fact that at the bottom of the mountain there is a greater gravitational pull from the Earth than at the top. After all, the bottom is nearer the Earth.

Again Einstein's calculations were right! The "clock" at the top of the mountain went slower than the one at the bottom. The differences were miniscule, but they were measurable. In 2010 this was confirmed using optical atomic clocks.

What all this means is that there is no one time in the universe, no one clock, no one calendar; different places are running at different speeds. Time is elastic.

Out in space, away from any strong gravitational tug, time flows quicker, but the nearer you get to a planet, or a star, the more time slows down.

And now for the real shock: if gravity gets really strong, like in a black hole where there is so much condensed stuff and gravity that not even light can get out, then a truly amazing thing happens. Time actually stops. This means that if you fell into a black hole the universe would have ended before you finished falling! For you time would have stopped while the universe continued on.

Feeling dizzy? You might like to go and splash your face again.

Mother Nature has some very deep and dark secrets. But the only thing more wonderful than the secrets is that we are beginning to understand them – and that you live in a time when they are being understood.

Wild, weird and wonderful scientific knowledge is exploding into our understanding all the time.

Scientifically speaking, there has never been a better time to live"!

Congratulations for being here now!

Could We Time Travel?

A startling question may be growing in the back of your mind at the moment. If time can be slowed down and if time can be speeded up, is it just possible that time could go backwards?

To put it another way, is it conceivable that we could travel back in time? Could we make a time machine? Some scientists have thought so.

One famous physicist, Paul Davies, who has been a professor of physics at several universities, has written a book called, *How to Build a Time Machine*. His basic conclusion seems to be that it might just be theoretically possible, but sadly not practically possible. However, he also says that that might not always be the case. He puts it in these words: "Just because time travel seems doubtful, or even impossible, to us today, doesn't mean that we can ignore its implications. It may be that easier ways to build a time machine will be discovered."

I suppose if some future civilization ever managed to make a time machine we would know about it as they would have visited us – unless of course they have, but keep themselves disguised.

Hurtling through Space

If all that stuff about time has made your head swirl, I now want to make it positively spin. I want you to glimpse something of what is happening to you at this very minute, right now, while you are reading this. This might make you feel more than a little nauseous.

You might want to go and get a bucket.

You are travelling at an enormous speed, faster than you could ever imagine. Let me give you just a glimpse of an idea of how fast you are hurling through space. If someone is standing at the equator of the earth then they are speeding round and around at just over 1,000 miles per hour (1670 km/hr).

It is easy to work this out. The Earth has a circumference of about 25,000 miles and it takes 24 hours to go all the way round while it is spinning. That adds up to just over 1,000 miles per hour. Obviously, as you go away from the equator and get nearer one of the poles you slow down because you travel less miles in one revolution, that is, in one day. At either pole you would be standing still, but you are turning a complete circle once every day. I hope you are with me so far. Now, that's just to warm us up.

If we were to do a similar calculation to work out how fast the earth is travelling around the sun as it goes on its yearly circuit then we get an answer of about 67,000 miles per hour (108,000 km/hr). So, from this you now know that at this very moment, right now, while reading this book, you are shooting through space at a breathtaking speed. But that's nothing. You are actually going far faster than that!

Far, far faster!

Our solar system, the sun and its group of planets, is part of a cluster of stars called the Milky Way. "Cluster" is not perhaps the best description of this group of stars; in fact there are some 200 to 400 billion stars in what we call our galaxy. This galaxy, the Milky Way, is a great swirl of stars, rather like what happens when cream is added to black coffee and given a stir. We are about two thirds out from the centre, and the whole thing is rotating in space.

Now for the really mind-bending stuff: It takes the Milky Way around 225 to 250 million years to make one revolution. In order to do that we are travelling at about 500,000 miles per hour – that's around half a million miles per hour! (About 800,000 km/hr) All these measurements are of course rather approximate.

Once you have got used to that, let me give you an even more unsettling thought. We now think that at the centre of most galaxies is a black hole and that the whole thing is rotating around it, held there by its gravitational pull. It's like being a fleck of paper swirling around the whirlpool of water going down a plug hole in the bath. Remember what happens if you fall into a black hole. Time stops. Let's hope we are going fast enough to keep our distance!

So, you are spinning around the earth at hundreds of miles per hour, which is spinning around the sun at around 67,000 miles per hour, which is spinning around what might well be a black hole at somewhere near a half a million miles per hour.

Dizzy?

You had better now close your eyes for a few minutes, before I begin to give your brain a rather *stretching* experience.

Big

Just how big is the universe? The short answer is big, very, very big. Let me try to give you an idea of just how utterly massive it is.

Suppose we got in one of our fastest rockets and wanted to travel only to the edge of our own solar system, how long would it take? Well, it would take about 12 years to get to Pluto, and that is nowhere near the edge. To get to the next nearest star, called Proxima Centauri, our spaceship would have to travel for . . . wait for it . . . twenty-five thousand years. In other words we are a long way from anywhere.

To imagine distances further than this we can no longer think in terms of our measly spaceships and their sluggish pace through space. We will now use the speed of light as a measurement. Light travels very fast indeed; in fact it travels at 300,000 kilometers per second. If light could bend around our earth it would travel seven times around it in one second. That is very fast. Light takes one and a quarter seconds to reach us from the moon, and it takes eight minutes to reach us from the sun. That means that the sunlight you now see outside your window left the sun eight minutes ago.

As the distances in space are so great the best measurement of distance is what we call the light year. A light year is the distance travelled by light in one year. Proxima Centauri is about 4.3 light years away. This means that when you look at this star the light you see left it 4.3 years ago. You are seeing what it looked like at just over four years ago. The star might be gone now, for all we know; it might have exploded. If it exploded today we would not see the explosion for 4.3 years.

To put it another way, when you look at this star you are looking back in time 4.3 years.

When the Hubble Space Telescope, a telescope that is on a satellite, views the Pleiades star cluster it is observing something 380 light years away. That means that the light took 380 years to reach us and that the sight seen happened 380 years ago.

The Veil Nebula is 1,400 light years away and what we see therefore occurred 1,400 years ago.

The nearest large galaxy to our Milky Way galaxy is called the Andromeda galaxy. It is about 2 million light years away and the light took 2 million years to reach us.

The nearest cluster of galaxies, called the Virgo Cluster, is about 50 million light-years away. So what we see happened 50 million light-years ago. We are looking a long way back into time.

The furthest sights observed in the universe are some 13-14 billion light year away.

The universe is very big, very, very big indeed.

Let us finish here with some frighteningly large statistics about our home in space, our galaxy, the Milky Way. It is about 100,000 light years across and is rotating like a great wheel in space. That means that travelling at the speed of light, a speed of 300,000 kilometers per second, it would take around 100,000 years to get from one side to the other. As we have seen, as this massive wheel of stars turns in space it takes between 225 -250 million years for it to make one rotation.

The Milky Way is unbelievably massive, it contains between 200 and 400 billion stars - and there are galaxies far bigger than ours out there. But the real jaw-dropper is that there are perhaps 200 billion galaxies in just the part of the universe that we can see.

The universe is simply enormous.

 Absolutely gargantuan!

And Getting Bigger

Space is not just shockingly big, what is really shocking is that it is getting bigger. Our universe is expanding. Our cosmos is growing – and very fast at that.

Most of the stars in the sky are moving away from us, and those furthest away are moving away the fastest. In fact, the furthest away are moving away at a rate not far from the speed of light. And the reality is even weirder than that: all these stars in the universe are moving away from each other. To each star it would appear that they are in the middle and everything is moving away from them. This needs some explaining.

Why doesn't it look like that when we stare out at the sky at night? Why do the stars not look like they are moving away? After all, they seem to drift across the dark sky in the same great pattern every night. The answer is that it's a bit like travelling on a train: those objects close by look like they are whizzing past very fast, but objects on the horizon hardly look like they are moving at all. Stars right out in space are so far away that they do not appear to be moving.

But if they don't look like they are moving, how do we know that they are moving? Basically, we know what the light coming from the stars should look like, but when we examine it carefully the wave lengths are different from what we would think they should be. The wavelengths are longer than we would expect them to be.

This is a well-known phenomenon of objects moving away from an observer. We have all experienced this with sound waves. When we hear a car sounding its horn drive past us it sounds higher pitched when it is coming towards us and lower pitched when it is going away – neee-ooow. As it is coming towards us the sound waves are squished up and so are higher pitched, but when it is moving away the waves are stretched out, making them sound deeper pitched. In a similar way, as the light waves coming from the stars have longer light wavelengths than we would expect, we know they are moving away from us. And by examining how much longer the wave lengths have become we can estimate the speed at which a particular star is receding from us.

That all sounds fine, until you think about it for a while. If everything is moving away from everything then it all came from a certain somewhere once. It looks like the whole universe exploded into existence – a sort of Big Bang. So the universe had a beginning; it had a start.

Now for the real mind-numbing point: if the universe back then was all crunched up into one tiny speck with massive gravity, then it was like the "Mother of all Black Holes". And as we know that in a Black Hole time stops, that would mean that there would be no time back then; time would not have started. The clock of the cosmos hadn't started.

And if that's confusing, here is something even more confusing: not only would there have been no time back then, there would also have been no space. Gravity not only slows and stops time; it also bends and shrinks space.

Then, right back then, at the very beginning, there was no time, no space and possibly no stuff. All that there was, was energy – and in a way we don't really understand.

So, as the explosion happened time was being created, space was being created – and stuff was being created.

Confused?

Hardly surprising; we are now looking at where the brightest minds fear to peer.

Patterns

What is truly stunning and mind-boggling is that our exploding cosmos is far from chaotic and random; it is actually very orderly; it is governed by laws, the laws of Nature.

Our universe and everything within it has a particular way of doing things. It is very well behaved, very well behaved indeed. It is as if Mother Nature has habits, habits that she just won't break. You would think that an explosion would end in a mighty mess, but that is far from the case. What we see is beauty, harmony and a degree of order that appears like it was designed.

One of the most amazing things about our universe is that it has certain patterns of behaving, certain secret laws, which if uncovered and understood can give us a degree of power over it. Knowledge is power.

If we discover, unravel and uncover these mysterious laws we can begin to control the universe, to manipulate the world around us to our advantage. Rather than being controlled by the forces and processes that buffet us and threaten to destroy us, we can begin to control them. Rather than being slaves, we can set out on a path to be masters. Instead of being subjects, we can rebel and become lords.

Humanity has only recently begun to do this, say for a few hundred years, but as we uncover more of what is behind Nature's patterns we are realizing that the horizons before us are endless. As we have said, the sky is nowhere near the limit.

Science is not so much about collecting and describing "things", it is much more about finding the mysterious and hidden connections between "things."

It is as we uncover and piece together the connections between "things" that we start to gain the power to control "things." It is only then that we can begin to learn to fly, to cure disease, to reduce famines, to be warned of earthquakes and tsunamis, or do whatever else is needed to ensure our survival.

The staggering outburst of knowledge and technology in recent times happened because just before that we began to realize something that is literally universe-shattering. We discovered an astonishing and staggering truth. We began to realize that such connections, or principles, or laws of Nature actually do exist. The universe is not controlled by magic or myth; the universe *is* governed by laws. It is only when people start believing that there is hidden treasure that they start seeking for hidden treasure. We now realize that the "hidden treasure" of Nature's laws really does exist. This is the Holy Grail of science.

We all experience "things" around us in our daily lives - and we rely on them to act in consistent ways. Gravity pulls down, sparks light fires, wind moves clothes on the washing line and friction produces warmth, and so on. Depending on this "sameness" is vital for us to carry out our daily routines and to survive and thrive. But for someone who is willing to dig deeper, there is more than a superficial "sameness" about the way things behave. As we prod and probe and poke we have begun to unveil principles that run through Nature, and the Laws that control her activities.

What is really astounding is that these principles or laws can be used, and what we do with them actually works.

The hidden code of the universe can be cracked, and once cracked it can be communicated - and it can be put to real work.

Once we have deciphered these secret laws imbedded in the fabric of the cosmos we can write them down, either in words, or in mathematical formulas, and then we can use them to actually do things.

This is no mere theoretical knowledge; the things we do with this knowledge actually do work. Planes fly through the air because we discovered secrets about air pressure and gravity and force, and when we communicated this knowledge to each other and then started to use it we found that it was not only accurate knowledge, but it was powerful knowledge.

Home, sweet home . . .

Our ball of rock spinning through space is our home. We call her Earth, but a better name for her would be Oceania or Aqua, as she is two thirds covered by water and all her life relies totally on this liquid. Without liquid water life is simply impossible.

Our planet feels safe and secure, nurturing and protecting life, but the reality is far more precarious. We constantly exist on the edge of destruction; we are always dangerously near extinction.

Our universe is very sterile – it's a dangerous place to live. Stars have temperatures of millions of degrees centigrade and issue radiation with unimaginable destructive force. There are places where gravity is so strong that hydrogen gas is turned into a silver solid, and other places where, as we have seen, the gravitational pull is so enormous that not even light can escape – the mysterious black holes. And then there is an unfathomable immensity of apparent nothingness – with temperatures below anything we can conceive. Space is utterly inhospitable and hostile to life.

Our planet protects us from the destructive forces of space and cocoons us in an environment that has just the right temperature, just the right atmosphere, just the right sunlight and just the right gravity to enable profoundly beautiful and complex life forms to flourish. Around us, above your head at this very moment, is a special force field known as the magnetosphere, which protects us from the destructive effects of the sun's rays. Without this we would all be dead. You would not be reading this and I would not have written it.

But this force-field has limitations: it can't protect us from bigger enemies.

Asteroids, great rocks in space, sometimes cross our path and just occasionally hit us. One the size of a house would wipe out a city.

There have been some near misses; in 1993 one sailed past us, missing us by only 145,000 kilometers, which in space terms is like having a bullet miss your head but ruffle your hair. That's a very near miss.

Around the world some massive craters have now been discovered that indicate colossal collisions in the past. One massive impact site off the shore near Mexico is 185 miles wide (300km) and was caused by a meteorite some 6 miles (10km) in diameter. This was difficult to find as it is a third of a mile (1km) beneath the surface, buried deep in rock. However, the crater is not the only evidence of this direct hit.

All around the world, in between certain layers of rock there is a thin slither, about 6 millimeters thick, of what might be called "space dust." What is strange about this layer is that it contains lots of a particular element that is very rare on Earth but much more abundant in space. This element, iridium, must have been sprayed all over the atmosphere of our planet when the meteorite exploded on impact and then gradually settled forming this extraterrestrial layer. Many think that it was perhaps this explosion that wiped out the dinosaurs and two thirds of all other species around the world as the deposit appears at just the right point in the rock layers. What is certain is that it would have produced phenomenal damage.

If one like this did slam into us right now it would hit us so fast that the air in the atmosphere underneath it would not be able to get out of the way in time and so would superheat – to a temperature 10 times the surface temperature of the sun. The meteorite itself would vaporize and blow a hole that would become the crater, decimating everything for miles, if not the whole world.

There would be world-wide fires caused by burning rocks falling from the sky as well as those massive waves called tsunamis, which could reach the height of skyscrapers. And as if this is not bad enough, it would probably set off a massive chain of earthquakes. But that is not all; then all the dust in the air would block out the sunlight, refrigerating the Earth and wiping out yet more life.

You wouldn't finish this book.

Don't Look Down

If you are feeling a bit queasy after learning how vulnerable you are to being instantly obliterated by a projectile from space, I now want you to look down and realize that from here on in things only get worse.

The middle of our planet, a mere 7,926.41 miles (12,756.32 kilometers) beneath your feet, is a nuclear reactor continually churning out unimaginable amounts of heat. Its temperature is between 4,000 and over 7,000 degrees centigrade, which is about as hot at the surface of the Sun.

This is very close to your feet. If you were to dig a hole to the centre of the Earth and drop your shoe down it, it would only take forty-five minutes or so to reach the middle (of course, in reality it would not keep falling as gravity would get less and less until your shoe just floated in mid-air in the middle of the planet — but the point I want you to grasp is simply this: this furnace is very close to you).

It is so hot down there that much of the rock between this central furnace and us is partially melted, being a sort of semi-solid that can flow. It is not very runny, only moving at a rate of about ten thousand times slower than the hour hand of a clock. Nevertheless, it still moves, the hot rock gradually rising and the cooler rock gradually sinking.

These circular currents of moving rock, called convection currents, are so powerful that they can actually move the earth's crust on the surface, shifting whole continents. Because of these currents great surface plates of the crust are gradually moving. This is why if you look at a map of the world it appears as if the continents once fitted

39

together. They did, and are moving apart at about the same speed that a finger nail grows.

These great sections of the moving Earth's surface are called tectonic plates. This is all fine until they suddenly grate or slip against each other in a kind of shudder. When that happens they produce a tremor that we call an earthquake, and if this happens under the sea the vibration causes a massive tsunami. Earthquakes and tsunamis kill tens of thousands of people. And if the molten rock beneath us ever bursts through a thin bit of the Earth's crust we have a volcano, which may have phenomenal destructive power - the ash and lava can bury whole cities and the clouds of ash can change the climate of the entire world. Our planet is an extremely dangerous place.

In August 1883, volcanic and earthquake activity on and around the Pacific island of Krakatoa was becoming more and more intense. And then on 26th August the main volcano blew out a cloud of ash 17 miles high into the sky, and ships as far as twelve miles away found hot rocks falling from the sky onto their decks. On 27th August explosions occurred that could be heard on islands in the Indian Ocean some 3,000 miles away, where they were mistaken for canon fire from nearby ships. The explosions were followed by tsunamis, which are believed to have been over 30 meters (100ft) high in places - that's three times as high as an average two storey house. The ash cloud ascended to a height of around 50 miles.

The shock waves were so phenomenal that they travelled around the globe seven times. These air pressure waves are thought to have caused waves on the sea, which travelled as far as the other side of the world, lapping the southern coast of England. After the explosion it was found that the island of Krakatoa had almost entirely disappeared.

In the year following the eruption, average temperatures around the world fell by more than 1 °C and the weather continued to be chaotic for years. The eruption injected an unusually large amount of a gas called sulfur dioxide high into the atmosphere, which was then transported all over the planet. This led to sunlight being reflected back into space, cooling our planet. All this changed the color of the sky, sometimes turning it blood-red.

When volcanoes exploded earlier in history they are thought to have so affected the weather that they caused world-wide droughts, famines and floods. Things like this have always occurred from time to time on our planet. Our home is very fragile, very fragile indeed.

Tiny

Having looked at our world, let us now see what it is made of at the tiniest level.

As the universe exploded into existence, energy formed into matter. There are several theories as to how this may have happened, but the really important point is that it can happen. We now know that pure energy can exist in a variety of forms, and one of those forms is matter. Indeed, we have learnt to do the opposite in practice, and so we can turn matter into heat and light energy; we do this in nuclear reactors and in nuclear bombs. There is a lot of energy squished and compressed into matter.

Matter, that is stuff, is made up out of atoms. Atoms are very small; in fact they are mind-bogglingly miniscule. We have already looked at the vastness of space, but we now need to grasp something about the smallness of atoms. Atoms are tiny, absolutely tiny; in fact a typical atom has a diameter only 0.00000008 centimeters. Now, that probably doesn't mean much to you, so we need to try a little experiment. This is a bit painful, but some sacrifices are needed in the cause of science. Put your hand on your head and carefully separate out one solitary hair. Now, quickly give it a sharp tug and pull it out.

Got it? It didn't hurt too much did it? Place this hair on a white sheet of paper and look at it and try to see how incredibly thin it is. How many atoms do you think there are across the width of this hair? The answer is about one million. Let yourself absorb that fact: there are about one million atoms standing as it were "shoulder to shoulder" to make up the diameter of that hair. Atoms really are very small indeed.

The Greeks first came up with the idea of an atom and the word really means uncuttable, or indivisible. That definition has stood the test of time well as it was only in the last century that we learnt that we could split them.

But what is in an atom? Atoms have a nucleus in the middle and things called electrons which whizz around it like planets circling around the sun. In reality, it is a bit more complicated that this, but this picture will do for the moment.

What is really strange is that atoms are mostly nothing but empty space. They are not very solid at all. If you took an orange and blew it up to the size of the earth, each atom would be about the size of a cherry. But you would not see anything; the nucleus would be too tiny for you to even glimpse it. You would have to blow that cherry up to the size of a football stadium, and then the nucleus would only be about the size of a bee. The electrons would be like specks of dust flying over the spectators' heads. Everything else is just space.

Let me put it another way, if the nucleus is about the size of a grape the electrons would be invisible specs of dust about a mile away. Again, everything else is empty space.

What this means is that *you* are almost entirely empty nothing; there is isn't anything solid about you. When it comes down to it *you* are mainly empty space. That's a little humiliating.

If that is true then why don't you fall through the chair you are sitting on? Or, if two people run into each other, why don't they just run straight through each other? The answer is that there are various forces between atoms that make us think that we are solid. It's a bit like holding two north poles of a magnet near each other and feeling them push each other away. The invisible force making them repel each other is like the forces between atoms that give us the illusion of things being solid. Things are not what they seem, not what they seem at all.

Outrageous

When we get down to the world smaller than an atom things start to get *really* weird, in fact so weird that no one quite grasps what is going on.

All those rules we talked about Mother Nature having start to break down and we have to throw them away and start to work out a whole new set. The trouble is that no one has been able to do that yet. This area of science has been given a special name all of its own, and that special name is called Quantum Theory. It is all so mind twisting that one famous scientist called Richard Feynman once famously said that, "If you think you understand Quantum theory, then you don't understand Quantum theory."

With that less-than-encouraging beginning let me try to give you a glimpse of the bizarre world that exists when you try to imagine things smaller than atoms. For a start, we now know that the electrons, which sort of whizz around the nucleus, have to go round on certain special tracks, like toy racing cars, or trains. Now here's the absolutely batty bit: the electrons can change tracks, but when they do so they do not pass through the space between. Confused? It gets worse.

We also now know that electrons sometimes appear to behave like particles and sometimes appear to behave like waves. A particle is something that exists at one place, whereas a wave is something all smeared out, existing in lots of places at once. This seems to be impossible. Surely electrons must be one or the other, either particles *or* waves! They can't be both! Are you more confused? Well, it gets even worse than that, in fact, much, much worse.

It seems that an electron as a particle does not exist until it is observed. Or, if I put it another slightly different way, if you look for an electron with an experiment designed to see a wave, you will see a wave. But then, if you conduct an experiment designed to see a particle you will see a particle. How you look at it changes what you see. You might want to go and take a cold shower at this point!

Before you give up the will to live with all this mind numbing stuff, let me pull one more trick out of the proverbial hat. Certain pairs of particles at this scale, that is, below the level of an atom, seem to "know" what each other are doing – even if they are separated by massive distances. These little particles have what is known as spin, and as soon as you determine the direction of the spin of one of the pair the other one will start spinning in the opposite direction, no matter how far away it is.

How does it know what its twin is doing? We don't really know. This appears utterly ridiculous. This is like there being two twins, one in Sydney and one in London. If the one in Sydney slips on a banana skin the other twin will do exactly the same at exactly the same time.

If your head feels as if it is about to explode, then you are not alone. When one of the scientists working all this out was asked how we might imagine an atom, he replied, "Don't try." Another said that the scientists had discovered "an area of the universe that our brains just aren't wired to understand." In other words, our brains are just not big enough. One other scientist said that scientists had coped with this problem "without thinking about it." It is all just too difficult – for the present, anyway. Mother Nature does not want to give up her secrets here. We are going to have to prod and poke her for quite a while yet – and perhaps a lot harder.

At the moment it is like we have two board games. For the game which deals with the larger normal universe we have found a reasonable set of rules that just about allows us to play the game. For the game that deals with the world tinier than atoms we have no clear rules and board is full of markings in some strange language. More than that, we have to play it in the dark while in a different room. We've got a lot to learn.

Now, why did I bother telling you all this, when everyone agrees it is just too difficult to understand? My reason is that this is Rebel Science and I want you to glimpse not only the easy stuff but the stuff at the frontiers of our knowledge. A century or two so ago, when Africa was not yet explored, you would be bored reading a description of a shopping trip to the local market, what you would want is a tale of an unexplored world, even if it was full of strange beasts and peculiar animals. My point is not to make you understand everything, for that would be impossible, but to make you feel excited that most things are still yet to be understood. Rebel science is real science. It's exciting stuff.

But let us now leave this miniscule wonderland and return to the safer and more comforting world of the atom.

Pass the Salt

If you took a chunk of any element, that is stuff made up of only one type of atom, and kept on cutting it in half, the smallest thing you could get down to and it still be that same element would be an atom.

If you divided the last atom you would no longer have that element. For example, if you kept chopping up a bar of copper you would get smaller and smaller chunks of copper. Once you got down to one atom of copper that is as small as you can go and it still be copper. If you break apart an atom of copper you no longer have copper.

What is really interesting about atoms is that they can team up with atoms of different elements in ways that totally change their characteristics. So, for example, if you were to get the metal sodium, which is soft like cheese and rather shiny when cut, and then place a chunk of it on your hand, it would burn its way right through, causing excruciating pain and poisoning your blood. And then, if you got hold of some of the green ghostly looking gas called chlorine and breathed in a great sniff of it, it might well be the last sniff you ever had. This gas was dropped on soldiers in the First World War to kill them in the trenches. Sodium and chlorine are toxic and deadly substances. However, if you react them together they turn into sodium chloride, which you add to your food to make it tasty. This is what we normally call salt. Sodium and chlorine have totally different characteristics from sodium chloride.

So, the really important point is this: when you react different elements together you end up with things that have totally different characteristics from what you started with.

Let me say that again in terms of atoms to make sure you have got the point. When you react different types of atoms together you end up with combinations of atoms bonded together that have utterly different properties from the atoms you started with.

Now, as there are over a hundred different elements, the numbers of combinations that are possible is immense. Scientists are still discovering new and exciting ones.

Chemists love reacting all sorts of atoms together to get totally novel and interesting substances. So, with different elements combining with each other in different ways, the potential number of substances that can result is enormous, if not infinite.

Before we move on I had better introduce a couple of terms that are really worth knowing. When different types of elements join together they are called **compounds**. When atoms bond together they are often called **molecules**.

Let me give you an example to make this clearer. Something like carbon dioxide is a compound made up of two elements, carbon and oxygen. Because it is made of more than one type of element chemically bonded together this is therefore a compound. It is also a molecule because there is more than one atom present.

However, to be a molecule the atoms *do not have to be* different types. So, for example, oxygen in the air around you is made of two oxygen atoms bonded together. It is therefore a molecule, but it is not a compound.

Now we have this under our belt we can get on to some *really* interesting stuff.

The Stuff of Life

All life is made of atoms, but the really mind-boggling question is, how do non-living atoms form living things like you and me?

Molecules, formed by atoms reacting together, have a vast array of different properties, totally unlike the individual atoms that they are made of. Now, some of those atoms can join "hands", or bond, to form quite, or very large molecules.

The really clever atom is carbon. It is clever because it can "hold hands" and make very long chains, sometimes with side branches. Each carbon atom has four "hands" and can therefore hold onto four other atoms. What we get therefore are long chains of carbon atoms, with various different types of atom hanging off the sides. You might well ask, what is so great about this? The answer is that if carbon couldn't form these chains you could not read this book. More than that, there would be no book – and more tragically, there would be no you. There would be no life. Life, certainly as we know it, is impossible without carbon atoms being able to "hold hands" with other carbon atoms to form chains. Carbon is the stuff of life.

Now for the utterly peculiar bit: if you get the "common old garden" elements carbon, oxygen, hydrogen, nitrogen and phosphorus, along with small amounts of things like potassium, calcium, magnesium and iodine, with a few others, you have pretty well all that you need to make – wait for it – life. Life is actually made of nothing special.

The amazing point is this: life has no magic ingredient, no supernatural essence. Life is simply normal common elements reacting with each other in particular ways. Very, very normal atoms join together to make complicated molecules that together, in vast numbers, react in such a special way that we call it living. Life is composed of chemical reactions. Life is not magic; it is only molecules, albeit complicated ones, bumping into each in certain ways.

The strangest, the most bizarre, the weirdest of all combinations of elements in the universe is when you get molecules that react with each other in such a way as to produce this stunning thing that we blandly call . . . life. Life is simply atoms and molecules reacting together in very complicated ways.

If you pick apart something living, say next door's cat, when you get down to the real essentials, all you have is a handful of elements. When it comes down to it, that is all the neighbor's cat actually is – and, if we want to be really honest, that is all you are. Again, it's all a bit humbling – if not humiliating.

Magic Molecules

What is the secret of life? When atoms are combined in a very special way, a strange type of molecule can be formed that can copy itself. *This* is the beginning of the miracle we call life. Self-replicating, or self-copying molecules are at the core of living things.

In these special self-copying molecules the carbons and other atoms in the long chain are very strongly bonded together, which means that there is a very strong long chain. This strong long chain has side branches all the way along its length, sticking out from it.

What I want you to do now is imagine two of these chains lying side by side. The side branches of one chain can stick weakly to the side branches of the other chain. This means the two chains can weakly stick together.

Have you got that?

Let me try to make it a bit clearer.

What you have got to get in your mind is a picture of two long strong chains that are weakly stuck to each other. It may help to think of a ladder with strong long sides but with steps that have a break in the middle of each of them, only held together by something weak like magnets. In essence, this is the magic molecule of life. Roughly speaking, this is what that special stuff called DNA is like.

What I want you to do now is some really hard thinking. Suppose we could pull the two sides of the ladder apart from the top, rather like a zip. Each of the steps would split in the middle, one by one.

Now, as the two sides of the "ladder" pull apart, other free floating "bits of ladder" (a step and a small part of a side) that are hanging around slot into the free spaces left at the end of "steps" that have opened up. Gradually, what happens is this: each side of the separated ladder joins up with these new free floating bits to form a new complete "ladder." You make two "ladders" out of the one original "ladder".

The really clever bit is this: the two new "ladders" are identical to the original "ladder." The "ladder" has duplicated itself; it has copied itself. This is the essence of life; this is the deep secret of life. This goes on at the heart of everything thing that we call living. Without this there is no life.

As I have already mentioned, the special molecule that does this self-copying is called DNA. DNA can copy itself. The letters DNA are just a short way of saying the long name of the molecule. The long name is deoxyribonucleic acid, which is quite a mouthful, but good to use if you want to impress someone.

Now DNA can also do something else other than copy itself. DNA can also code for making another type of molecule called proteins. Proteins are the workers that carry out the chemical reactions that keep us alive.

Each side of the "steps" of the ladder is made of one of four chemical "letters." These letters can be arranged in different orders to code for the making of a wide range of proteins. Each long strand of code for each protein is called a gene. So DNA not only copies itself, it also codes for the making of the chemical "workers" that keep us alive.

Your body is made up of trillions of little 'bags' called cells, and in the nucleus of almost every one of them is this code to make you, you.

You Could Go to Prison for This!

Now let's talk about something that is totally illegal, at the moment anyway.

If you took the nucleus out of one of your cells and put it in a woman's egg that had had its nucleus removed, and then give this cell a tiny electric shock and implanted it in her womb, or any woman's womb, the baby that would grow would be your identical twin. This would be your clone. Theoretically at least, you could make millions of you-look-alikes walking around the planet.

Good idea?

It is illegal to do this with humans; after all, it is a bit Frankensteinish. Although, it would not surprise me if some multi-millionaire has already paid for it on the Black Market!

There was a film in the 1960s where mad Nazi scientists got some of the cells of Adolf Hitler just after he died, kept them in a deep freeze, and then later cloned him lots of times in Brazil. The idea was to generate a new leader for the Fourth Reich, which would rise up and capture Europe, if not the world.

This sort of cloning has already been done with sheep – and people will pay to have it done to their pet cat, or slug, or something.

The point is this: the code to make the whole of your body like it is, is pretty well in every one of your cells. Each of your cells contains the information to make you. Amazing!

There is Far, Far More to You than Meets the Eye

There is more to your DNA than this duplicating itself and coding for worker proteins. Your DNA is not only clever; it is also long – very long indeed.

DNA is an incredibly long molecule when it is unfolded. Normally DNA is super-coiled up, rather like a ball of string. However, if you managed to unravel it you would find that it is extraordinarily lengthy.

If you added together all the lengths of all the DNA molecules in all of the cells in an average adult, the length would be as long as the distance to the sun and back, not once, not twice, but seventy times.

You didn't know you had it in you!

But before you get too cocky, there is something else you need to know about your DNA – and you might not find this so wonderful.

All Life is Connected

I don't want to be insulting, but you share about half of your DNA with a banana. We have some pretty strange cousins! Our closest relative is the chimpanzee, which shares about 98 percent of our DNA.

Our DNA is unique to us, but it is similar to other animals and plants. Although no other living thing has DNA identical to ours, we do have great similarities.

What is truly amazing is that living things can read each other's DNA code. A bacterial cell can understand and decode human DNA; a horse can understand crab DNA; a fly can understand mouse DNA. We all speak the same genetic language.

The more you think about this the weirder it becomes. Think of something really ugly or horrible, something that scares you, like a spider, or snake, or slug. If you put your DNA into the nucleus of one of its cells it would understand it and decode it and make what your DNA commanded it to make. And the same is true in reverse; your cells could decode spider, or snake, or slug DNA. And it does not have to even be an animal; your cells can read stinging nettle DNA, and slime mould DNA can read yours. We can all read each other's genes. Every living cell on Planet Earth speaks the same biological language.

That's awesome!

This fact is used in medicine. When people suffer from the disease called diabetes they can't control the sugar levels in their blood because they are unable to produce a special chemical, a hormone, called insulin. Scientists can cut out the DNA from human cells that codes for making this substance and then stick it into the DNA of some bacteria. What now happens is staggering.

The bacterium reads the human gene as if it is its own and starts making insulin. In fact the bacterium also divides, multiplying itself, making more and more cells with the human insulin DNA within it. In the end there are lots of these bacterial cells with the human insulin gene spliced into them, and so together they produce lots of insulin, which we use to inject into people with diabetes.

Now for the really bizarre stuff: suppose a gene coding for making an eye was taken out of some mouse cells and put into the cells of a fly larva that would later develop to become a leg.

What would happen?

It sounds like something out of a horror movie – but it has been done, and what happened surprised everyone. The fly did develop an extra eye, yes, an eye on its leg – but the eye was not a mouse eye; it was a fly eye. The fly cells not only read the mouse DNA; they used it to trigger the making of their own type of eye.

This is incredible.

Life is full of mystery.

Intrigued?

Addicted yet?

Bags of Life

Although it might be different on some other planet out there far off in space, life, as we know it, is composed of little microscopic bags called cells. Cells are the basic building blocks of living things.

Our DNA is packaged in a tiny parcel in the middle of our cells that is called the nucleus. And outside of this nucleus is a special area where most of the chemical reactions of life take place, controlled by proteins. This working area is called the cytoplasm. And all of this is wrapped up in a "skin" of fatty molecules called the cell membrane. It is a bit like a kitchen where the cookery book is the nucleus and the writing DNA, where the worktops are the cytoplasm and where the kitchen walls are the cell membrane.

When you put all this together you have the nucleus in the middle, which copies itself and passes out plans for making protein workers to the cytoplasm. Outside the nucleus you have this cytoplasm where the protein workers are made and do their work, and then around all that is this special skin-like cell membrane. Together, the nucleus, the cytoplasm and the cell membrane make up the basic unit of life. We call this the cell. It is the bag of life.

To use another picture, the nucleus is like the office of a factory where plans for making things are copied. The cytoplasm is like the workshops of the factory where things are made. The cell membrane is like the wall around the factory where things are carefully let in and out.

The basic building block of life is the cell. Every living thing on Planet Earth is made of cells; plants are made of cells; animals are made of cells; you are made of cells. These cells are little bags, and as we have seen, they have a controlling nucleus, a chemical workshop of cytoplasm and the outer bag itself, the cell membrane. Your body has in the order of 10,000 trillion of them, all working away to keep you alive.

These cells of yours burn food gently and slowly to give you the energy you need; they make all your chemicals and keep them at just the right levels; they attack intruding bacteria; they make you read; they make you think. And this all happens without you having to do anything consciously – it's all automatic.

What is more, all the time they are continually being replaced as they wear out – most only last a month or so. Brain cells are different; they last as long as you last. Well, that's not quite true. You got about a hundred billion of them when you were born and that's all you will ever get – but the bad news is that some five hundred or so of them die each hour. At this moment your brain is dying. Depressing!

So, you are made of trillions of these microscopic living bags, all specialized to do different things – and all working together in harmony. Just look in a mirror. You are amazing!

But the real shock is that there was a time when you were just one cell. Let me say that again – and read this slowly: there was a time when you were a single cell, an individual solitary cell with a nucleus, cytoplasm and a cell membrane.

Now, before that humiliates you too much, let me tell you that that cell was unique. It was totally original. There has never been a cell like that before – and there never will again, however long the universe exists. Let me explain.

Once upon a time an absolutely tiny sperm cell from your father bumped into an absolutely massive egg cell from your mother – in fact the egg cell was 85,000 times bigger. That was no equal match!

It had been a race between millions of sperm cells, but once the one that was going to form part of you hit the egg, a special capsule in its head let out a powerful chemical that dissolved part of the egg cell's membrane. The nucleus of the sperm then entered. At this point the egg cell did something remarkable: it sent a signal all over its membrane to stop any other sperm from injecting in its nucleus. The gates were closed. All that was needed to code for you was present.

You were on your way.

Your journey of life had begun.

This solitary new cell then divided into two, then that two each split, making four, and so on. After forty seven or so divisions there were roughly the 10,000 trillion of them that makes you, you.

What is really clever is that all these cells, as they divide, know what to become, and very rarely get it wrong. Some become brain cells, some bone, some eye, some muscle and so on – and they all become what they become in the right place. This is incredible.

All the time these dividing cells were chemically talking to each other, making sure that everything is going according to plan. And it worked. You have one nose, two eyes, one heart, one head – and so on. Just think what would happen if a few cells decided to rebel and do their own thing and make another leg, an extra foot – or a reserve head under your armpit! On second thoughts, let's not think about it!

Joining Forces

All these cells coming together to make you is a pretty clever trick, in fact many cells in Nature don't bother joining forces like this; they go it alone.

The majority of living things are single celled. Many of the things that cause us disease, like bacteria, are simple single cells, as are the bacteria that rot down our waste material. Single-celled living things thrive all around us and are very successful.

Bacteria can divide staggeringly fast. In fact, if a single bacterial cell like one in your stomach kept dividing in ideal conditions, after 44 hours the growth would equal the weight of the Earth. It is really good that bacteria don't often get ideal conditions in your guts – or anywhere for that matter! Single-celled living things are a force to be reckoned with.

There are some extremely strange life forms on our planet that are somewhere in between single-celled and multi-celled. There is a type of green slime mould made up of individual plant cells that on very special occasions does something that could have been made up for a grotesque tale of horror – and this will make the blood drain from your face.

What happens is this. The single green slime cells all start moving towards each other, slowly coming together to produce what can only really be called a slug – and it can move; it then crawls to an open lit space to reproduce. When was the last time you cleaned that damp corner under your bed? You wouldn't want something to crawl out and reproduce on your face!

Before we continue, I need to tell you something else about this slime mould slug made of lots of individual cells that come together to move. What I am about to tell you is utterly remarkable.

This slug-like thing can "remember!"

If researchers pass dry air over it, it will slow its crawling down. If they do this once every hour it will slow down each hour. Now for the shocking bit: if the scientists stop blowing dry air over it, it will still slow down at the time the air should have been blown over it. Somehow it "remembered." And then, if the air was not blown over it for several hours the mould would still slow down in an hourly cycle for about three hours. It would then stop cycle of slowing down.

Now it gets even more amazing: if the blowing was started again just on one occasion, the slug would start its hourly cycle of slowing down each hour. It had clearly "remembered" what had happened before and started up its hourly cycle of slowing down.

Just exactly how does it do this? You might well ask how something with no brain, even no nerves, can "remember" anything? Well, to be honest we are not sure at all. Research is continuing.

Anyway, getting back to where we were before, there are some living things that are single cells, and there are some, like the slime mould "slug" where the cells can sometimes come together and work as a team. And then, on top of this, there are some strange animals that are made of lots of cells but will not die if they are pulled apart. If you take an animal like a sponge, for example, and chop it up and ram it through a kitchen sieve into a pot of water, the cells will all come back together again to reform the complete animal. You couldn't do that!

However, most of the things that we actually see around us are plants and animals that are many-celled, and prefer to stay that way. They can't be chopped up and then somehow grow back together again. Their cells must stay together; they perform as permanent teams. We call these multi-cellular living things – and the things they get up to are utterly astonishing.

Bodies

When cells team up and form bodies they can do some amazingly stunning things. Prepare to be shocked.

A cheetah can run at a speed of 70 – 75 mph (112 – 120 km/h); a whale can hold its breath for up to two hours; a killer whale can swim at a speed of 30 mph (48.4 km/h).

If anything, birds are even more astonishing. There is the Ruppell's Griffon Vulture, which can fly to height of 37,000 ft (11,277m) – that's over 7 miles or 11 kilometers up in the sky!

And then there is the bird called the Arctic tern, which migrates from its northern breeding grounds near the Arctic all the way down to the oceans around Antarctica and back again, a round trip of about 44,300 miles (70,900km).

Let that fact settle into your brain: this bird migrates from near the top of the Earth, all the way to near the bottom and then back again.

On top of this, the Peregrine Falcon can dive at a speed of 124 mph (200 km/h).

If all that does not shock you, some humming birds can beat their wings at the incredible speed of 90 beats per second. That's too fast to see!

When cells work together as bodies the results are just staggering.

Your own body is pretty amazing too.

Let us think about your heart for a moment. From very early on in your mother's womb it started beating and will continue until . . . well, generally as long as the rest of you continues. Your heart pumps 343 liters of blood around your body each hour, which adds up to some 3 million liters in a year. It is hard to imagine what that means, but it is enough fluid to fill four large swimming pools.

And did you know that your thigh bones are stronger than concrete? They need to be considering the amount of walking you will do. In the average lifetime the normal person will walk the equivalent of five times around the equator – that is some 125,000 miles (about 200,000 km).

Your leg muscles do an incredible amount of work, but these are not your strongest muscles. Relative to size, the strongest muscle in your body is – wait for it - your tongue. And to finish these encouraging statistics on how utterly amazing you are, consider this: throughout you life you will grow roughly 98 feet (30m) of eyelash hair. Your body is very clever. Give yourself a pat on the head.

But before you get too cocky there is a more creepy fact about our bodies that you need to know – and this is not that pleasant at all.

The Walking Dead

If you look at your teacher you are looking at what is dead. Everything you can see about her is dead – except, eerily, her eyes.

Her eyes are the only living things you can see – and they peer out of utter death and decay. All you can see of her – or yourself, if you look in a mirror, is deceased, dead, departed. I had better explain this before you run away screaming (although you might find it difficult to run away from yourself!).

The surface cells of our skin slowly die as they are replaced by newer living ones coming up beneath. That top layer of dead skin is continually flaking off and blowing away on the draughts until it settles – as dust. Much of the dust around your home is the dead remains of all your family – and you. You are continually breathing in the dead leftovers of people. No wonder dust makes us sneeze!

So, all our skin on the surface of our bodies is dead, every inch of it has expired – it is deceased. In fact, the average adult carries around with them about 2kg of dead skin. And if you find that repulsive, there's worse to come, much worse.

All these dead cells seem very unpleasant to us, but for some tiny animals they are a gastronomic feast. These flakes of skin and all the oil that continually oozes out of you is delectable to the creatures that live in your bed with you.

Your mattress may have a living population of some two million of these things called mites, munching and nibbling their way through your bodily secretions. I suppose the worst thing I have to tell you in this regard concerns where you rest your head at night. If you pillow is about six years old it may have forty thousand of these little creatures gobbling away all around your face. More than that, it has been calculated that ten per cent, that is one tenth, of the weight of your pillow may be made up of dead skin, living mites, the rotting bodies of dead mites and of course mite poo.

Sleep well!

Perhaps you really should change that pillow.

Super Teams

Bodies are not the end of the story; they can sometimes join together to form something even greater.

So, atoms make molecules, and some very special molecules can carry out reactions that we call life. These take place in little bags called cells. The majority of living things are individual cells. However, cells can work together and grow together to form what we call bodies. Most of the living things we see every day are multi-cellular like this; that is why they are big enough for us to see them. But it does not stop here; living bodies tend to come together with their own kind to work in teams, which almost exist as bodies themselves.

Take ant nests for example. Ants can live together in colonies which are highly organized and consist of millions of individuals. There can be ants that are totally dedicated to finding food, others that are given the job of protecting the nest, still others that care for the young – and then there is a queen who does all the reproducing with a few males. It is as if the whole nest exists as one body with each type of ant operating as a sort of organ. The level of organization is phenomenal. In fact, some scientists call the colonies super-organisms. Ant colonies exist like one large living thing.

The individual ants can communicate with each other with special chemicals called pheromones. More than this, it has been found that ants are the only group of animals, apart from mammals, to actually show each other how to do things, that is, to actually interactively teach each other. Together, ants can work out and solve complicated problems. For example, some species of ant are able form chains to bridge gaps over water or through gaps between plants. Others also make floating rafts that help them survive floods. This all appears very intelligent.

A crushed ant gives off an alarm pheromone chemical that causes nearby ants into a sort of battle-frenzy, which then attracts reinforcements from elsewhere. It is like an alarm system or battle-cry. Some ant species even use "propaganda pheromones" to confuse enemy ants and make them fight each other.

Ants attack others and defend themselves by biting and, in many species, by stinging, often injecting or spraying pain causing chemicals like formic acid. Trap-jaw ants have jaws which snap shut faster than any other predators within the animal kingdom. Some species have been shown to snap their jaws shut at a speed of between 78 to 143mph (126-230 km/h). There would be no escape from that!

Other ants can use their jaws like catapults to throw out intruders or fling themselves backwards to escape attack. Another type of ant has a special gland near its mouth which has a sac that contains a poison that can paralyze attackers. When the nest is attacked the ant breaks this sac and squirts the poison over the invaders. The defending ant itself then dies. It has sacrificed itself to defend its nest. This is not the only example of ants committing suicide in defending their colony. In other species, a small group of ants leave the nest before night and seal it from the outside, leaving themselves vulnerable to predators, and thus are generally eaten.

Some species of ants attack and actually invade and take over neighboring ant colonies. Others invade rival colonies to steal eggs or larvae, which they either eat or raise as slaves. Just think about that: ants will raise other ants to be their slaves. They are almost as bad as humans.

Extreme types of slave-making ants are unable to feed themselves and need captured slaves to work for them so that they can survive.

Other ants take their queen and fight their way into the nest, or even use pheromones to confuse the defending ants into tricking *them* to carry their queen in. That is really cheeky.

The queen of certain African ants allows herself to be dragged by other ants into their nest. What she then does is truly grotesque and utterly horrific. She climbs on the back of the resident queen and then bites off her head. The resident workers do not seem to notice this, and she then lays her eggs and the whole nest acts as slaves to raise *her* young. One particularly type of queen does not even bother to cut the defending queen's head off herself. She secretes a chemical that causes that queen's young to cut their own mother's head off. She does not even do her own murderous dirty work.

Several types of ants can actually act like farmers looking after herds of cows. What they do is care for another insect called an aphid. The aphid sucks sap out of plants and produces a rich fluid, rather like a sort of milk, called honeydew, which the ants harvest and take back to their nests. The ants even protect their flocks from predators. In some cases the aphids secrete the honeydew in response to the ants tapping them with their antennae. It is like the aphids are milked by the ants. The ants in turn not only keep predators away, but will even take the aphids with them when they move to another area.

Some ant species capture and herd caterpillars, which are massively bigger than the ants. They lead the caterpillars to special feeding areas in the daytime, and then even bring them inside the ants' nest at night for safety. The caterpillars have a special gland that secretes honeydew when the ants massage them.

All these strategies make ants very successful; they are found in almost every part of the planet apart from Antarctica, and are thought to weigh as much as all the people on Earth! (It is difficult to do an experiment to check this exactly!)

The Living Planet

We have been on a mind-reeling journey, all the way from the utter immensity of space, down to the bizarre world within the atom. And from there have pulled back the lens and observed atoms joining to make molecules, and molecules reacting in near-miraculous ways to form life's building blocks called cells. We have seen cells clinging together to form bodies, and now we have grasped that even bodies can so work together as a unit to form a weird life-form called the colony. But that is not the end. There is more.

Here, again, we go beyond school textbook science. Back in the 1970's a scientist who worked for the North American Space Agency (NASA) came up with a radical new theory. He believed our planet is alive. His name is Dr James Lovelock.

At first sight the theory is strange and unbelievable, but when looked at closer it is not as weird as it sounds. He is not saying that the Earth can think, or anything like that. What he is saying, though, is that when looked at as a whole, the Earth is itself nearly as complicated as life, with millions of processes all balancing and interacting and controlling each other. The whole thing works to sustain life like you and me because all these processes work together as a whole, rather like the reactions in a cell, or the organs of our bodies.

The Earth "feeds" on the sun using all its plants, which can capture the sun's energy with that special green chemical that plants have, called chlorophyll. This capturing of the sun's energy is a process that scientists call photosynthesis. This turns the light energy into chemical energy.

Now, all this chemical energy is available to other living things like animals, which eat the plants. And some animals eat the animals that eat the plants. So, all energy comes from the sun. (Well, that is what we are often told at school, but strictly speaking it is not true. We now know that there are some bacteria that can get their energy from the chemicals made by volcanoes deep under the sea. However, it is true to say that by far, most energy for life comes from the sun – in fact, *almost* all of it.) The energy from the sun flows down from plant to animal – and everything is then fed on by bacteria and fungi as they rot it all. This flow of energy sustains life.

But there is more. Nothing is wasted; elements used by living things, like carbon and nitrogen, are reused again and again; they are recycled.

Perhaps the most stunning example is carbon, which is extracted from the gas in the atmosphere called carbon dioxide using the sun's rays in photosynthesis, and built into life sustaining chemicals containing lots of energy. The carbon is then released back into the atmosphere by living things burning up that chemical energy in a process we call respiration. Some carbon is of course released from animal and plant life by real burning. This is all finely balanced.

Even rocks are recycled. There is that massive nuclear reactor at the centre of the Earth that heats up the semi-fluid rock deep underground called magma and sets up rising and falling convection currents that move whole continents. At the edges of the great moving plates of the Earth's crust some old rock gets sucked down and melted, whereas elsewhere other new molten rock is pushed up, cools and sets. In this way rock is continually renewed and replaced.

And then on top of all this, our planet even seems to have a circulatory system. The Earth transports temperature and substances in special great flowing currents of air and water all around its surface. Nature is wonderfully balanced.

The Earth can also protect itself. The upper atmosphere acts as a special chemical "skin" that can keep out dangerous rays from the sun – and this is helped by a "force field" called the magnetosphere, which deflects some harmful rays back out into space.

All this is quite stunning. Our planet may not actually be alive, but it certainly has a complexity that has many similarities to being alive.

Are You a Disease?

The vital question is, of course, if the Earth is sort of "alive", can it die?

Well, yes. If a meteorite hit it like we thought about earlier, then it could wipe out all these processes, including all living things. We know that it has been hit various times in the past but has always recovered – so far.

What about disease? Can the earth become diseased? Think about some of the diseases that affect us. Cancer is when some type of cell starts dividing out of control and damages and even destroys the rest of the body. Could it be that we humans are becoming like that? Are we the Earth's cancer?

Our population is enlarging out of control; we are using up the Earth's resources; we are poisoning our planet with pollution. Perhaps, from the planet's point of view, the billions of *Homo sapiens* teeming over its surface is a disease.

However, we are not like cancer cells in that we can think. We are conscious beings. It is not to our advantage to destroy our own home, and working together we can make decisions to stop it happening. Indeed, the greatest difference between us and all other life is that we can do just this: we can make conscious decisions. We are aware of ourselves; we can think; we can reason; yes, we can make *conscious* decisions.

Our Latin species name, *Homo sapiens*, actually means Wise Man. Time will tell whether we are wise enough to save ourselves.

Life that Thinks

Everyone agrees that we humans are the best at thinking in the animal kingdom, but perhaps it is not really fair to say that we are the only ones that can think at all.

Some animals not only appear to be able to think in some way, but even show the ability to care for others. It is as if they have some kind of rudimentary moral awareness.

Dolphins have been found to protect people swimming in the sea and will even circle around them to keep sharks away. More than that, whales have been found helping other whales that are drowning, pushing them to the surface so that they can breathe.

Once, when an elephant was shot, others were seen coming around her and trying to help her to keep standing. When she went down on her knees they tried to lift her up with their tusks, and after she finally collapsed they attempted to push food into her mouth to revive her. When they finally realized she had died they covered her body with soil and branches. All this shows some kind of conscious awareness, even if limited.

Chimpanzees have been spotted going into water to save another chimp from drowning, and even sacrificing their own life to do so. And they also have a sense of fairness, getting angry if one is fed more than another. They can even throw temper tantrums in such a situation. Less pleasantly, they will work in well-organized gangs to go through the jungle and attack other chimpanzees, trying to invade their territory. It looks as if they are waging war.

Some of the great apes use simple tools, like a stone, to crack open nuts. Others have been found to perform certain actions that appear very like rituals, particularly when they see something that looks like it is filling them with a sense of awe, like a great waterfall, or a thunderstorm.

The Lord of the Thinkers

However, having said all this, it is us; it is human beings who are at the pinnacle of consciousness and intelligence. What we can do is to think about thinking; we can think about ourselves; we are self-conscious.

With the power of language and the ability to think, we can reason and learn from the world around us, digging out its hidden mysteries. And once we have discovered its secrets, we can communicate these in words to each other. This stops us forgetting what we have learnt and it means that we can build on each other's discoveries and not repeat each other's mistakes.

This takes us back to where we started. We began our tour of the universe by realizing that there was a way of poking Mother Nature so that she revealed her secrets to us – and that we could, gradually, bit by bit, begin to control her. Knowledge is power.

Hopefully, we can learn to harness that knowledge in a way that uses but not abuses the world around us.

Perhaps, just perhaps, we really can become *Homo sapiens* – that is, wise man.

What part will you play in this?

To Inspire You . . .

Aerodynamically, the bumble bee shouldn't be able to fly, but the bumble bee doesn't know it so it goes on flying anyway.

Mary Kay Ash

Every great advance in science has issued from a new audacity of imagination.

John Dewey

Facts are the air of scientists. Without them you can never fly.

Linus Pauling

From now on we live in a world where man has walked on the Moon. It's not a miracle; we just decided to go.

Tom Hanks

If an elderly but distinguished scientist says that something is possible, he is almost certainly right; but if he says that it is impossible, he is very probably wrong.

Arthur C. Clarke

No amount of experimentation can ever prove me right; a single experiment can prove me wrong.

Albert Einstein

Nothing has such power to broaden the mind as the ability to investigate systematically and truly all that comes under thy observation in life.

Marcus Aurelius

Nothing in education is so astonishing as the amount of ignorance it accumulates in the form of inert facts.

Henry B. Adams

Nothing in the universe can travel at the speed of light, they say, forgetful of the shadow's speed.

Howard Nemerov

Only two things are infinite, the universe and human stupidity, and I'm not sure about the former.

Albert Einstein

People think of the inventor as a screwball, but no one ever asks the inventor what he thinks of other people.

Charles Kettering

Science does not know its debt to imagination.

Ralph Waldo Emerson

Science has made us gods even before we are worthy of being men.

Jean Rostand

The great tragedy of science – the slaying of a beautiful hypothesis by an ugly fact.

Thomas Huxley

The most exciting phrase to hear in science, the one that heralds new discoveries, is not 'Eureka!' but 'That's funny...'

Isaac Asimov

There was no "before" the beginning of our universe, because once upon a time there was no time.

John D. Barrow

Touch a scientist and you touch a child.

Ray Bradbury

What is a scientist after all? It is a curious man looking through a keyhole, the keyhole of nature, trying to know what's going on.

Jacques Yves Cousteau

Your theory is crazy, but it's not crazy enough to be true.

Niels Bohr

Science is a way of thinking much more than it is a body of knowledge.

Carl Sagan

Science is organized common sense where many a beautiful theory was killed by an ugly fact.

Thomas Huxley

Science is simply common sense at its best, that is, rigidly accurate in observation, and merciless to fallacy in logic.

Thomas Huxley

People think of the inventor as a screwball, but no one ever asks the inventor what he thinks of other people.

Charles Kettering

And Now For Something Completely Different!

We are sometimes told that fact is stranger than fiction, but here we discover that fact is *funnier* than fiction – often far funnier. Normal life certainly has its hilarious side; it is simply a case of noticing it. Comedy is just beneath the surface of our everyday routines.

Jack Renniks learnt to appreciate this when, as a freelance columnist for a British newspaper, he had to come up with a humorous article every week. Gradually, he started to see the funny side of life everywhere, the absurd, the ridiculous and yet always the truth. In his books, **A Bunch of Laughs**, there are insights that will certainly make you smile, if not laugh out aloud. Here is something to whet your appetite.

Have you heard of the Bristol poo man? That's the man, David Carlisle, who was sentenced to four years in jail for stealing two bags of dog poo.

Just in case you haven't caught up on this one, he was caught after brandishing a knife at one Marion Budd, while she was out walking her dog in the park.

"Give me your bags" he demanded to the shocked lady. And so, good citizen, Ms Budd, who had been contentiously scooping up after her dog and tying its feces into bags, did as she was requested. Carlisle was later arrested and paid rather a high price for a bit of dog mess.

At a party some of us were chatting about this story and it sparked off our memories of similar "wrapped goody" tales we had heard. One lady, an Australian, told what had happened to someone she had known as a girl, back "down under." There is a train station near Sydney with an incredibly long flight of stairs. An elderly lady was struggling up this one day with a very large suitcase, when a man came along and offered to help her. However, he was not honest, nor sincere, and it was a ruse to steal her bag and all its contents. She had been on her way to visit her son in the country; when she got there she relayed her sad tale to him and he was obviously very angry with the man and very sorry for her.

'Never mind,' she said, 'it doesn't really matter at all, in fact he did me a favor. Last week, my pet dog had died in my flat in the city and I had nowhere to bury it, so I thought I would bring it to your house to lay to rest in your garden.'

www.ingramcontent.com/pod-product-compliance
Lightning Source LLC
Chambersburg PA
CBHW051347170526
45166CB00002B/997